Fact Factory

From Atoms to Zebras (And Everything in Between)

H.X. Erickson

Sweet Pea Press

Published by Sweet Pea Press

www.sweetpeapress.net

First Edition 2025

Book Cover and Illustrations by Clayton Crooks

ISBN Paperback: 978-1-967211-01-2

ISBN Digital: 978-1-967211-00-5

To the next generation of explorers, scientists, and dreamers – the ones who know that the strangest things in life are often the most fascinating. Keep asking questions. Keep being surprised.

Contents

Chapter 1

CREATURES GREAT AND SMALL

This chapter is all about the amazing creatures that share our planet, from the biggest blue whales to the tiniest insects, the prehistoric dinosaurs to today's playful pets. Whether they're fierce predators, like sharks and eagles, or friendly giants, like elephants and giraffes, every creature has something incredible to share. Get ready to dive into the wild world of nature and be amazed by these fun, surprising, and downright unbelievable facts!

Amazing Animals

The heart of a blue whale is as big as a small car. Imagine needing a parking spot for that!

Octopuses have three hearts and blue blood. They'd make great royalty with all that blue blood!

Male seahorses carry and give birth to their babies. Talk about a great dad!

Elephants can "hear" with their feet, detecting vibrations through the ground. Talk about having a "good ear" for music—except it's their feet!

Butterflies can taste with their feet. Imagine eating pizza by stepping on it—gross but fascinating!

Axolotls can regrow entire limbs and even parts of their heart or brain. They're like real-life superheroes with a built-in "fix-it" power!

A panda's diet is 99% bamboo, but they sometimes eat small animals. Guess even pandas like a little variety in their snacks.

The mantis shrimp punches with the force of a bullet. Never challenge this shrimp to an arm-wrestling contest!

An ostrich's eye is bigger than its brain. At least they've got their priorities straight—see first, think later!

Kangaroos can't walk backward. They're all about forward momentum—no time for regrets!

Dolphins sleep with one eye open to stay alert for predators. Now that's what I call multitasking!

Platypuses are one of the few mammals that lay eggs. They're the ultimate animal mash-up—part duck, part beaver, part who-knows-what!

Dazzling Dinosaurs

The Argentinosaurus is believed to be the largest dinosaur, weighing up to 100 tons. That's like stacking 14 elephants on top of each other—talk about heavy traffic!

Velociraptors were only about the size of a turkey, not as large as in movies. Thanksgiving would've been wild back then!

The Stegosaurus had a brain the size of a walnut. It wasn't winning any spelling bees, that's for sure.

Tyrannosaurus rex could crush bones with its powerful jaw. Good thing they weren't picky eaters—they'd just crunch through it all!

The name "dinosaur" means "terrible lizard." Not all of them were scary, though—some were more like clumsy lizards.

Some dinosaurs, like the Microraptor, had feathers and could glide. Imagine a chicken taking flight—it's a little like that!

Ankylosaurs had armored bodies and a clubbed tail for defense. They were basically prehistoric tanks!

Triceratops had three horns and could weigh as much as a school bus. That's one bus you wouldn't want to cut off in traffic!

The first dinosaur fossil was discovered over 200 years ago. Can you imagine the surprise when they thought they'd found dragon bones?

Pterosaurs, flying reptiles, were not technically dinosaurs. They were like the nosy neighbors of the dinosaur world, always flying overhead.

The Brachiosaurus could reach leaves up to 40 feet high. A giraffe would need a ladder just to keep up!

Spinosaurus is the largest known carnivorous dinosaur. Think T. rex was scary? Spinosaurus was even bigger and loved to swim!

Some dinosaurs swallowed rocks to help grind their food. Who needs teeth when you've got a built-in blender in your belly?

Dinosaur fossils have been found on every continent, including Antarctica. Even dinosaurs liked to chill in the snow!

Dinosaurs lived for over 165 million years. That's way longer than humans—dinos were the ultimate long-term residents!

The Oviraptor's name means "egg thief," but it likely protected its own eggs. It got blamed for crimes it didn't commit—poor Oviraptor!

The largest dinosaur eggs were about the size of a football. Imagine finding that in your backyard Easter egg hunt!

Deinonychus inspired the design of the "raptors" in Jurassic Park. Hollywood made them bigger, but the claws were just as scary!

The colors of dinosaur skin remain a mystery to scientists. Were they green, striped, or pink polka-dotted? Nobody knows!

Birds are modern-day descendants of theropod dinosaurs. Next time you see a pigeon, think of it as a tiny T. rex in disguise!

Intriguing Insects

Ants can lift objects up to 50 times their own weight. That's like you lifting a car—time to hit the gym, humans!

Bees communicate by dancing to show the location of flowers. Imagine giving directions by busting out your best dance moves!

A dragonfly can fly backward as easily as forward. They'd be amazing helicopter pilots.

Some butterflies drink the tears of animals for salt. Crying isn't always a bad thing—it's snack time for butterflies!

Fireflies produce light without generating heat. They're nature's coolest little flashlights.

Termites are more closely related to cockroaches than ants. Who knew these tiny builders were part of the roach family reunion?

A cockroach can live for weeks without its head. That's one way to keep life going... until hunger kicks in.

Mosquitoes are considered the deadliest animals because of the diseases they spread. Tiny but mighty—and not in a good way!

Ladybugs can eat up to 5,000 aphids in their lifetime. They're the ultimate pest control squad!

The Goliath beetle can grow up to 4.3 inches long. Imagine finding that in your backyard—yikes!

Crickets can estimate temperature based on their chirps. Nature's little meteorologists are on the job!

There are over 12,000 species of ants worldwide. Ants are basically everywhere, even if you don't invite them.

Cicadas can remain underground for up to 17 years before emerging. They're like nature's ultimate hibernators.

The monarch butterfly migrates up to 3,000 miles annually. They're tiny travelers with serious frequent flyer miles.

Male moths can smell a female from over 7 miles away. Talk about a strong sense of smell—no perfume needed!

Some spiders use their silk as a parachute to travel long distances. They're adventurers in their very own hot air balloon rides.

Leafcutter ants use leaves to grow their own food – fungus. They're not just gardeners—they're farmers, too!

Praying mantises can turn their heads 180 degrees. They'd crush it in a game of peek-a-boo.

The Hercules beetle can carry 850 times its own weight. That's an incredible display of strength in the insect kingdom!

Mosquitoes are attracted to the carbon dioxide in your breath. So next time, try holding your breath around them—just kidding!

Brilliant Birds

Hummingbirds can flap their wings up to 80 times per second. That's like running a marathon every time they move—no wonder they need so much sugar!

Ostriches are the fastest running birds, reaching speeds of up to 45 mph. Forget racehorses—ostriches are the real track stars.

The albatross can glide for hours without flapping its wings. It's like the chillest airplane in the animal kingdom.

Penguins can drink seawater thanks to a special gland that filters out salt. They'd never need a water purifier on a camping trip!

The harpy eagle has talons bigger than a bear's claws. They'd win every arm-wrestling contest—if they had arms!

Some parrots can live up to 80 years. That's a lifetime of repeating everything you say—choose your words wisely!

A group of crows is called a "murder." That's one spooky-sounding bird squad.

Woodpeckers can peck up to 20 times per second without hurting their brain. They'd be unbeatable at headbanging contests.

The Arctic tern migrates over 40,000 miles each year. These birds are the ultimate long-distance travelers—move over, jet-setters!

Flamingos get their pink color from the food they eat, like shrimp. So, if they ate broccoli, would they turn green?

The kakapo, a flightless parrot, is one of the rarest birds in the world. It's a bird so rare, it could have its own reality show.

Pigeons can recognize themselves in a mirror. They're not just flying rats—they're self-aware flying rats!

Owls can rotate their heads 270 degrees. They'd be great at spying on neighbors without getting caught.

The kiwi lays eggs that are nearly 20% of its body weight. That's a huge amount!

Peacocks' tail feathers can be over 6 feet long. They're the ultimate show-offs at nature's fashion show.

The peregrine falcon is the fastest bird, diving at speeds of over 240 mph. That's faster than a sports car in freefall!

Ravens are known to use tools and solve puzzles. They're like the MacGyvers of the bird world.

Seagulls can drink both fresh and salt water. They'd thrive on a deserted island—minus the coconuts.

The lyrebird can mimic almost any sound, including chainsaws. Imagine a bird pulling pranks with fake chainsaw noises—terrifying and hilarious!

Puffins use their beaks to carry multiple fish at once. They're nature's multi-tasking fishermen!

Super Sharks

Great white sharks can detect a drop of blood in the water from three miles away. That's like finding a needle in a haystack—underwater!

The whale shark is the largest fish in the world, growing up to 40 feet long. Imagine swimming next to a bus—that's basically a whale shark!

Hammerhead sharks have 360-degree vision thanks to their unique head shape. They'd be unbeatable in hide-and-seek.

Sharks have been around for over 400 million years, predating dinosaurs. They're basically the great-grandparents of the animal kingdom.

Some sharks can glow in the dark to camouflage themselves. They're like swimming glow sticks at a shark party.

Tiger sharks are known as "garbage eaters" because they eat almost anything. If it fits in their mouth, it's on the menu!

The basking shark has a mouth nearly 3 feet wide but only eats plankton. Proof that big mouths don't always mean big appetites.

Sharks' teeth are constantly replaced, with some replacing up to 50,000 in a lifetime. Talk about the ultimate dental plan!

Sharks can sense electrical signals from other animals. They have a built-in sixth sense that helps them detect unseen movement in the water.

The smallest shark, the dwarf lantern shark, is only about 8 inches long. It’s like the chihuahua of the shark world.

Nurse sharks can "walk" along the ocean floor using their fins. Who needs legs when you’ve got fins that double as feet?

Shark skin is made of tiny teeth-like structures called dermal denticles. That’s why it feels like sandpaper—ouch!

Greenland sharks can live for over 400 years, making them the longest-living vertebrates. They'd have some serious stories to tell—if they could talk.

Sharks don't have bones – their skeletons are made of cartilage. They're basically swimming noses!

Lemon sharks are named for their yellowish color. They're not sour, but they sure are zesty!

Bull sharks can swim in both salt and freshwater. They're like the amphibians of the shark world.

Goblin sharks have a long snout and extendable jaws. They look like something straight out of a sci-fi movie.

Mako sharks are the fastest sharks, reaching speeds up to 60 mph. They'd leave most speedboats in the dust—er, bubbles.

Megalodon, an extinct shark species, had teeth as big as a human hand. Imagine brushing those chompers!

Sharks' noses can detect prey hiding under the sand. They're the ultimate treasure hunters of the sea.

Chapter 2

EARTH'S AMAZING WONDERS

In this chapter, we explore the breathtaking wonders of our planet, from the depths of the oceans to the icy peaks of glaciers, and everything in between. Discover how rainforests create their own weather, why volcanoes erupt, and how glaciers shape the land over time. From dazzling Northern Lights to the power of tectonic plates, Earth is full of incredible surprises just waiting to be uncovered. Get ready to journey through the natural marvels of our world with these fascinating and awe-inspiring facts!

Outstanding Oceans

The ocean covers over 70% of Earth's surface. It's like the ultimate swimming pool—but way deeper!

The deepest part of the ocean, the Mariana Trench, is over 36,000 feet deep. That's deeper than Mount Everest is tall—talk about a hidden world!

Some jellyfish can glow in the dark. They're like underwater nightlights!

Octopuses can squeeze into spaces as small as their beak. Forget keys—just send an octopus to unlock your door!

Sea otters hold hands while they sleep to stay together. That's the cutest sleepover ever!

A group of dolphins is called a "pod." It sounds like they're all listening to the same playlist.

The Great Barrier Reef is the largest living structure on Earth. It's like an underwater city, and the fish are its citizens!

Some crabs decorate their shells with sponges and other sea creatures. They're nature's original fashion designers.

Whale songs can be heard for miles underwater. It's the ultimate ocean concert!

Starfish don't have brains or blood. They must really know how to keep things simple!

The fastest fish, the black marlin, can swim up to 82 mph. Forget jet skis—just hitch a ride on this guy!

A shrimp's heart is located in its head. That's what I call "thinking with your heart!"

The ocean contains 97% of Earth's water. That's a lot of salty drinks!

Sea turtles can hold their breath for up to 5 hours. That's longer than I can sit through a movie!

Clownfish live in sea anemones, which protect them. It's like having a built-in security system for your house!

The vampire squid uses bioluminescence to confuse predators. Nature's way of saying, "Lights out!"

Manta rays have a wingspan of up to 29 feet. They're like the graceful gliders of the sea.

Sharks have been swimming in the ocean for over 400 million years. They're basically the grandparents of the sea.

Coral reefs protect coastlines from strong waves. They're like the ocean's shield.

There are underwater lakes and rivers. Water inside water—how cool is that?

Radiant Rainforests

Rainforests produce 20% of the world's oxygen. Every breath you take might be thanks to a tree!

More than half of the world's species live in rainforests. It's the ultimate biodiversity hotspot!

The Amazon rainforest is so vast that it creates its own weather. Imagine a forest with its own forecast channel!

The poison dart frog's bright colors warn predators to stay away. It's like nature's "do not touch" sign.

Some rainforest trees can grow over 200 feet tall. That's taller than a 20-story building—what a view!

Sloths move so slowly that algae grow on their fur. They're like walking greenhouses!

Rainforests can receive over 200 inches of rain annually. It's like living under a never-ending sprinkler!

Orangutans make umbrellas out of leaves to stay dry. Who knew they had DIY skills?

The rainforest floor is so dark that plants grow giant leaves to catch sunlight. It's survival of the biggest leaf!

Rainforest plants are used to make medicines. Nature's pharmacy is open 24/7!

The harpy eagle can carry prey as heavy as monkeys. It's like a flying strongman!

Bromeliads, a type of plant, hold water in their leaves, creating mini ponds for frogs and insects. That's one fancy apartment complex!

Some ants in rainforests act like farmers, growing fungus to eat. Even ants have day jobs!

Rainforests have existed for over 50 million years. They're ancient history that's still alive!

Toucan beaks are big but super light because they're mostly hollow. They're flashy and functional!

Some butterflies in the rainforest pretend to be poisonous by copying the wing patterns of toxic species. It's nature's version of a really convincing disguise!

Jaguar spots help them blend into the rainforest shadows. Nature's camouflage is on point!

The rainforest canopy is so thick that rain can take 10 minutes to reach the ground. It's like a natural umbrella!

Leafcutter ants can carry leaves 50 times their weight. That's the ultimate grocery haul!

Rainforest rivers are home to pink dolphins. Who needs unicorns when you've got pink dolphins?

Perplexing Plants

Bamboo is the fastest-growing plant, growing up to 35 inches a day. It's a true speed champion of the plant world!

The Venus flytrap eats bugs for extra nutrients. Plants that snack? Now that's cool!

Rafflesia, the largest flower in the world, smells like rotting meat. It's not winning any fragrance awards!

Cacti can store up to 200 gallons of water. They're the ultimate desert survival experts.

The corpse flower blooms once every few years and smells terrible. Nature's way of saying, "Don't come near me!"

Mangrove trees can grow in saltwater and help prevent coastal erosion. They're like nature's barrier reefs on land.

Sunflowers follow the sun as it moves across the sky. They're like little solar-powered plants!

Bananas are technically berries, and banana plants are giant herbs. Bananas are full of surprises!

Baobab trees can store thousands of gallons of water in their trunks. They're like giant tree water bottles!

The giant sequoia is one of the tallest and oldest trees on Earth. It's like a time machine made of wood!

Pitcher plants trap insects in their cup-shaped leaves. They're like the rainforest's bug-zappers.

Some plants can grow without soil, using only air and water. Floating plants? Sign me up!

Wolffia, the smallest flowering plant, is about the size of a grain of rice. Tiny but mighty!

The cacao tree produces the seeds that become chocolate. Thank a tree for your favorite treat!

Some mushrooms glow in the dark. They're like little disco lights in the forest!

The Joshua tree can live for hundreds of years. It's the desert's ancient guardian.

Dandelions spread their seeds using the wind. Make a wish and let nature do the rest!

Eucalyptus trees are koalas' favorite snack. Koalas are nature's picky eaters!

The strangler fig tree wraps around other trees to grow. It's like a plant hug—except not so friendly.

Lotus flowers can survive for thousands of years. They're the ultimate survivors of the plant world!

Glorious Glaciers

Glaciers cover about 10% of Earth's surface. That's like a giant ice cube blanket for the planet!

The largest glacier, the Lambert Glacier in Antarctica, is over 60 miles wide. It's wider than some countries!

Glaciers can move, but very slowly—up to 50 feet per day. It's like they're on a super slow-motion road trip.

Blue ice in glaciers gets its color because it's so dense it absorbs all light except blue. Nature's coolest shade!

Some glaciers create deep cracks called crevasses, which can be hundreds of feet deep. Careful where you step—it's a long way down!

Glaciers are found on every continent except Australia. Even Africa has glaciers on Mount Kilimanjaro!

When glaciers melt, they leave behind unique landscapes like fjords and valleys. It's like nature's way of redecorating.

Glaciers store 70% of the world's freshwater. They're the planet's frozen water bank!

Icebergs form when chunks of glaciers break off into the sea. Titanic-sized, literally!

Some glaciers are over 30,000 years old. They've been chilling out since the Ice Age—literally!

The sounds of cracking ice in glaciers can be heard for miles. It's like nature playing an icy drum solo.

The Perito Moreno Glacier in Argentina is one of the few glaciers still growing. A glacier that's bucking the trend!

Polar bears often travel on floating chunks of glaciers. It's like their icy taxi service.

Glacier ice can trap ancient air bubbles, giving scientists a peek at Earth's past. It's like nature's time capsule!

Glaciers carve mountains and create sharp peaks. Ever wonder why some mountains look so jagged? Thank a glacier!

Glaciers act as giant mirrors, reflecting sunlight back into space. They're like the planet's sunglasses.

The term "glacier" comes from the French word for ice. Fancy name for a giant block of frozen water!

Melting glaciers contribute to rising sea levels. It's like filling a bathtub too full!

Alaska has more than 100,000 glaciers. That's a lot of icy neighbors!

Glaciers move by sliding on a thin layer of water underneath. Ice skating on a massive scale!

Energetic Earth

Earthquakes happen when tectonic plates shift. It's like the planet cracking its knuckles!

The Ring of Fire is a zone around the Pacific Ocean with lots of volcanoes and earthquakes. Sounds dramatic, right?

Volcanoes erupt when pressure builds up inside Earth's crust. It's like a soda bottle that's been shaken too much!

Mount Everest is the tallest mountain on Earth, standing at 29,032 feet. It's taller than most planes fly!

The Earth's core is as hot as the surface of the sun. Talk about a fiery center!

Geysers, like Old Faithful, erupt when underground water is heated by magma. Nature's version of a tea kettle!

Lava can reach temperatures of over 2,000°F. Definitely not something to touch!

The Grand Canyon was carved by the Colorado River over millions of years. Water really knows how to sculpt!

The Northern Lights are caused by charged particles from the sun hitting Earth's atmosphere. It’s like the sky’s private light show.

The Earth’s crust is broken into 15 major tectonic plates. It’s like a giant puzzle we all live on.

Sinkholes form when the ground collapses into underground caves. Nature sometimes pulls the rug out—literally!

The Sahara Desert is so large it’s roughly the size of the United States. That’s a lot of sand!

The Mariana Trench is the deepest part of the ocean. It’s deeper than Mount Everest is tall—mind-blowing!

Earth’s magnetic field protects us from harmful solar radiation. It’s like the planet’s invisible shield.

The Andes Mountains are the longest mountain range in the world. A hike that long would take forever!

Hot springs occur when groundwater is heated by Earth's magma. It's nature's hot tub—just don't forget your towel!

Landslides happen when gravity pulls loose rocks and soil downhill. It's like nature's big cleanup—messy but effective.

Volcanic ash can travel thousands of miles in the air. It's like a smoky cloud that takes a world tour.

Earthquakes can create tsunamis when they happen underwater. A big wave with a lot of power!

Some caves are formed by acid rain dissolving limestone. Who knew rain could dig holes?

Chapter 3

THE FORCES THAT SHAPE OUR WORLD

In this chapter, we dive into the incredible forces that shape the world around us and keep it in motion. From the gravity that keeps us grounded to the friction that slows us down, forces are the invisible powers behind everything we see, feel, and experience. Discover how forces like buoyancy keep massive ships afloat, how tension holds things together, and how lift helps planes soar through the skies. These fantastic forces are all around us, making the ordinary truly extraordinary. Get ready to explore the science that powers our universe in this action-packed chapter!

Cool Chemistry

Water can boil and freeze at the same time under special conditions. It's like water can't decide what to do!

Diamonds are made of carbon, the same element in pencils. Who knew your pencil had so much sparkle potential?

The human body is about 60% water. You're basically a walking water balloon!

Dry ice isn't ice—it's frozen carbon dioxide. And it creates spooky fog for Halloween!

When you mix baking soda and vinegar, they fizz because they release carbon dioxide gas. It's like a volcano in your kitchen!

Your stomach acid is strong enough to dissolve metal. Luckily, it only works on food (and not spoons)!

Chemicals in fireworks create different colors. That's why your July 4th is so colorful!

Salt can lower the freezing point of water. That's why it's sprinkled on icy roads—magic science at work!

Soap works by breaking up grease and dirt into tiny pieces. It's the superhero of cleaning!

There are more atoms in a glass of water than there are stars in the galaxy. Who knew water was so crowded?

Helium makes your voice sound funny because it's much lighter than air. It's the science of squeaky voices!

The smell of rain comes from oils released by plants. Science even explains that fresh rainy smell.

Liquid nitrogen is so cold it can shatter a rubber ball. Freeze and smash—it's like magic!

The bubbles in soda come from carbon dioxide gas. Your drink is fizzy science in a bottle.

Chemical reactions power glow sticks. That's why they light up your dance party!

Some metals, like sodium, explode in water. Definitely not the kind of thing you'd want in your pool!

Acids and bases react to create salt and water. They're opposites that make peace in the end.

Chlorophyll makes plants green and helps them make food. It's like plants have their own built-in solar panels.

You can dissolve an entire egg's shell in vinegar. Science says goodbye to hard eggshells!

A single lightning bolt has enough energy to power 100 lightbulbs for a day. That's one electric zap!

Powerful Physics

Gravity is the force that pulls us toward Earth. Without it, we'd all be floating like astronauts!

Light travels at a speed of about 186,000 miles per second. Talk about fast—it could circle Earth 7.5 times in one second!

An object in motion stays in motion unless acted upon by another force. That's why a rolling ball eventually stops—it's Newton's fault!

Sound travels faster through water than air. So maybe whales hear music better than we do!

Electricity flows because of tiny particles called electrons. They're like little runners powering your gadgets.

A rainbow forms when sunlight bends through water droplets. Nature's most colorful science project!

Magnifying glasses work by bending light to make things appear bigger. Science really zooms in on the details!

The center of gravity is why you can't bend forward without falling if your back touches a wall. Try it—it's science proving you wrong!

Energy can't be created or destroyed—it just changes forms. It's like energy is the ultimate recycler!

Objects weigh less on the moon because it has less gravity. Space is the ultimate weight-loss plan!

A rocket launches because of Newton's third law: For every action, there's an equal and opposite reaction. That's a lot of science just to go to space!

A pendulum keeps swinging because of gravity and inertia. It's like a perfect science dance.

Friction slows down moving objects. That's why sliding on ice feels so different from sliding on grass!

The force of air resistance explains why a feather falls slower than a rock. Thanks, science, for giving us cool experiments to watch!

An echo happens when sound waves bounce back to you. Nature's way of saying "I heard you!"

Hot air rises because it's less dense than cold air. That's how hot air balloons take flight—science lifts them up!

The pressure inside a soccer ball helps it bounce. Without it, your ball would just flop around.

If you drop two objects of different weights in a vacuum, they fall at the same speed. Even a bowling ball and a feather—mind-blowing!

Gyroscopes work by spinning and staying balanced. That's why they keep bikes and airplanes steady!

A black hole is a point where gravity is so strong that even light can't escape. It's the ultimate cosmic vacuum cleaner.

Magnetic Mysteries

Magnets always have two poles: north and south. No matter how you slice them, they'll always stick together!

Earth acts like a giant magnet, with its own magnetic field. That's why compasses point north—it's Earth's way of giving directions.

Magnets can attract or repel each other depending on their poles. Opposites really do attract!

Electromagnets are magnets powered by electricity. Switch them on, and they're magnetic magic!

The strongest magnets in the world are made of rare-earth metals. They're small but mighty!

Magnetite is a natural magnetic rock. It's like Earth's original magnet.

Some animals, like pigeons and sea turtles, use Earth's magnetic field to navigate. They've got built-in GPS!

Magnets don't work on all metals. Gold and silver? No dice—but iron? Totally!

Magnets lose their power if they're heated too much. Even magnets need to chill out!

A magnetic field is invisible but can be seen using iron filings. Science makes the unseen visible!

MRI machines use powerful magnets to look inside your body. It's like science peeking inside without opening you up!

Maglev trains float above the tracks using magnets. That's how they go so fast—no wheels needed!

The sun has its own magnetic field. That's one powerful solar magnet!

Magnets were discovered over 2,000 years ago in ancient Greece. They've been attracting attention ever since!

Junkyards use giant electromagnets to lift cars. They're like the bodybuilders of the magnet world!

Magnets can be used to store data in computers. Who knew your laptop had magnetic magic inside?

The Northern Lights are partly caused by Earth's magnetic field. It's a light show with a magnetic twist!

Magnets can make metal objects float if the magnetic force is strong enough. It's like a sci-fi hoverboard!

The Earth's magnetic poles have flipped many times in history. That's a real "pole reversal"!

Compass needles are tiny magnets that align with Earth's magnetic field. They're like mini scientists pointing the way!

Funky Fossils

Fossils are the preserved remains or traces of ancient organisms. They're like nature's time capsules!

The word "fossil" comes from the Latin word meaning "dug up." It's the perfect name for buried treasures.

Most fossils form in sedimentary rock. That's rock made from layers of sand, mud, and time!

Amber fossils are formed from tree sap that traps bugs or plants. It's nature's way of making jewelry with history inside.

The oldest known fossils are about 3.5 billion years old. That's older than your great-great-great-everything!

Dinosaur footprints can become fossils, showing how they walked. It's like finding ancient muddy sneakers!

The largest dinosaur fossil ever found belonged to an Argentinosaurus. Its thigh bone alone was taller than most people!

Petrified wood forms when wood is replaced with minerals. It's like turning a tree into stone art.

Fossils are often discovered after erosion uncovers them. Time and weather are the world's sneaky paleontologists!

Sharks' teeth are some of the most common fossils. Sharks have been losing teeth for millions of years—it's like they were preparing for us to collect them!

Fossilized dung, called coprolite, helps scientists learn about ancient diets. Yes, even prehistoric poop can be useful!

Trilobites, an extinct sea creature, are some of the most common fossils. They're the fossil celebrities of the ocean.

Fossils show that some dinosaurs had feathers. Who knew T. rex might have looked a little fluffy?

Some fossils form as molds, leaving an impression of the organism. It's like a natural stamp!

A "living fossil" is a species that's still alive but looks almost the same as its ancient relatives. Crocodiles are like dinosaurs that never left!

Coal is made from fossilized plants that lived millions of years ago. Your barbecue might be powered by ancient ferns!

Fossilized raindrops have been found, showing patterns of prehistoric weather. Even rain leaves a mark on history!

Fossil hunters use tools like brushes and chisels to gently uncover bones. It's like a treasure hunt with extra patience.

The Burgess Shale in Canada is famous for its perfectly preserved fossils of ancient sea creatures. It's like the ultimate fossil museum!

Scientists can sometimes extract ancient DNA from fossils. It's like nature's version of Jurassic Park—but way safer!

Fantastic Forces

Gravity is what keeps us grounded on Earth. Without it, we'd be floating around like astronauts in space!

The stronger the gravity, the more something weighs. So technically, you'd weigh less on the moon—who needs a diet plan?

Friction is the force that slows things down when they rub together. That's why ice is slippery—there's barely any friction!

The force of air resistance slows things down as they fall. That's why parachutes work—it's like the air is catching you.

Buoyancy is the force that helps things float in water. That's why even heavy ships don't sink—thank you, buoyancy!

Centripetal force keeps objects moving in a circle. Without it, carnival rides would fling you into space—yikes!

Electromagnetic force helps keep atoms together. It's like the invisible glue of the universe!

Tension is the force in ropes or strings when they're pulled tight. Think of it as a tug-of-war with physics.

Elastic force happens when objects like rubber bands stretch and snap back. It's fun until you get snapped in the face!

A force is measured in units called newtons. Yep, named after the apple-and-gravity guy!

Magnets create a force field that can pull or push objects. It's like magic, but totally science!

Inertia is the resistance of an object to change its motion. That's why a rolling ball keeps rolling unless something stops it.

Pressure is the force applied over an area. That's why sharp knives cut better—they focus the force!

Action and reaction forces are always equal and opposite. That's why rockets launch into space—they push down, and the Earth pushes back!

Spring force makes a spring return to its original shape after being stretched or compressed. It's why trampolines are so bouncy!

Lift is the force that helps planes stay in the air. It's basically the secret to flight.

Drag is the force that slows objects moving through a fluid, like air or water. That's why birds and planes are designed to be sleek.

Torque is the force that makes things rotate. It's why wrenches are so good at twisting bolts.

Thrust is the force that moves objects forward. Rockets, jets, and race cars owe their speed to thrust!

Every force has both a size and a direction. It's like giving physics a GPS for how to act!

Chapter 4

EPIC STORIES OF THE PAST

Step into the past and uncover the incredible stories, events, and people that shaped our world. From the towering pyramids of Egypt to the Great Wall of China, from daring heroes like Joan of Arc to visionary inventors like the Wright brothers, history is packed with epic achievements and unforgettable moments. In this chapter, you'll travel through time to explore the fascinating landmarks, jaw-dropping inventions, and heroic figures who made history come alive. Get ready to be amazed by the power of the past!

Historic Happenings

The Great Wall of China is over 13,000 miles long. That's one wall you definitely don't want to walk the entire length of!

The first Olympic Games were held in 776 BCE in ancient Greece. Back then, they didn't even have gold medals—just olive wreaths!

The Titanic took over 2 years to build but sank in less than 3 hours. Talk about a dramatic twist!

The Wright brothers' first flight lasted only 12 seconds. Short but sweet—now look where airplanes are!

Mount Vesuvius erupted in 79 CE, burying the city of Pompeii. It was a disaster, but it left behind a time capsule of Roman life.

The Berlin Wall divided a city for 28 years before coming down in 1989. It's a powerful symbol of change and unity.

The Rosetta Stone helped unlock the secrets of ancient Egyptian writing. It's like the ultimate cheat code for hieroglyphs!

Neil Armstrong became the first person to walk on the moon in 1969. His footprint is still there—no wind to blow it away!

The printing press was invented by Johannes Gutenberg in 1440. Without it, we might not even have books today!

The first known writing system, cuneiform, was developed around 3200 BCE. It looks like little triangles—imagine texting in that!

The Eiffel Tower was originally meant to be temporary. Good thing they kept it—it's now one of the most iconic landmarks!

Stonehenge has been standing for over 4,000 years. And no one really knows how or why it was built—spooky!

The first modern Olympics included events like tug-of-war. Now that's a sport we'd love to see again!

Anne Frank wrote her famous diary while hiding during World War II. Her words have inspired millions around the world.

Cleopatra lived closer in time to the moon landing than to the building of the pyramids. Time is wild!

The Great Fire of London in 1666 destroyed much of the city—but it also led to the creation of modern fire safety codes. A disaster that reshaped history!

The Statue of Liberty was a gift from France to the U.S. in 1886. It's the biggest "thank you" gift ever!

The Vikings were the first Europeans to reach North America—500 years before Columbus. They beat him to it, but without the fanfare!

The golden age of pirates lasted only about 30 years. But they packed a lot of swashbuckling into those few decades!

Albert Einstein's brain was preserved after his death for scientific study. Even after life, he was still a genius!

Heroic History

Joan of Arc led an army at age 17. Most teens are busy with homework, but she was busy saving France!

Harriet Tubman helped hundreds of slaves escape through the Underground Railroad. She was called "Moses" for leading her people to freedom.

Martin Luther King Jr. gave his "I Have a Dream" speech in 1963. His words still echo around the world today.

Leonardo da Vinci not only painted the Mona Lisa but also invented early designs for planes and submarines. Talk about a multitasker!

Mahatma Gandhi led peaceful protests that helped India gain independence. He showed that change doesn't always need violence.

Amelia Earhart was the first woman to fly solo across the Atlantic Ocean. She made history soaring above the clouds!

Florence Nightingale revolutionized nursing during the Crimean War. She was called "The Lady with the Lamp" for helping soldiers at night.

Abraham Lincoln signed the Emancipation Proclamation in 1863. He literally changed the course of history with his signature!

Helen Keller, despite being deaf and blind, became an author and activist. She proved there are no limits to what you can achieve.

Marie Curie was the first woman to win a Nobel Prize—and she won two! Talk about double the brilliance!

Rosa Parks refused to give up her bus seat in 1955. That small act sparked a big movement.

Alexander the Great built one of the largest empires in history by the age of 30. Talk about ambitious!

Susan B. Anthony fought for women's right to vote. She believed everyone's voice should be heard!

Nelson Mandela spent 27 years in prison fighting against apartheid. His resilience led to a new era of freedom.

George Washington became the first U.S. President. He literally wrote the book on being a leader!

Anne Sullivan, Helen Keller's teacher, taught her to communicate using touch. She unlocked a whole world for Helen.

The Wright brothers dreamed of flying and made it a reality. Their dream became everyone's reality!

Frida Kahlo turned her pain into powerful paintings. Her art is still inspiring people today.

Neil Armstrong's moonwalk inspired generations to look up at the stars. His "small step" was a giant leap for everyone.

Harriet Beecher Stowe's book, Uncle Tom's Cabin, fueled the fight against slavery. Proof that words can change the world!

Time-Traveling Facts

A day on Earth used to be only 18 hours long. The planet's been slowing down—lazy Earth!

Sharks have been around longer than trees. The ocean wins in the race for oldest residents!

The first humans appeared about 200,000 years ago. Compared to Earth's age, we're the new kids on the block!

Dinosaurs lived for over 165 million years. That's way longer than humans have been here—respect!

The Ice Age covered much of the Earth in glaciers. Imagine snow days lasting thousands of years!

The Great Pyramid of Giza was the tallest structure in the world for 4,000 years. Move over skyscrapers!

Earth is 4.5 billion years old. And it still looks pretty good for its age!

The continents were once one giant landmass called Pangaea. It's like Earth started with one big jigsaw puzzle!

Some stars we see today burned out millions of years ago. We're looking at ancient starlight!

The Roman Empire lasted over 1,000 years. That's a lot of history packed into one civilization!

The first written language appeared around 5,000 years ago. Before that, people just had to talk—or draw pictures!

The Bronze Age got its name from people using bronze tools and weapons. Imagine naming a whole time period after your favorite tool!

The Industrial Revolution changed how people worked and lived. It was the start of machines doing all the heavy lifting.

The oldest known city, Jericho, dates back over 9,000 years. Talk about prime real estate!

Fossils show that whales used to have legs. Imagine a whale walking into the ocean one day and never looking back!

Electricity wasn't commonly used until the late 1800s. Before that, people just went to bed when it got dark!

The universe is about 13.8 billion years old. It's hard to even imagine that many candles on a cake!

The first clocks were powered by water, not batteries. Who needs AA batteries when you have water power?

The oldest shoe ever found is over 5,500 years old. And no, it wasn't a fashionable sneaker!

The calendar we use today was introduced by Julius Caesar. Thanks, Caesar, for giving us a schedule!

Peculiar Planets

Mercury is the smallest planet in our solar system. It's like the baby sibling of the planetary family!

Venus spins in the opposite direction of most planets. It's doing its own thing—planetary rebellion!

Earth is the only planet with liquid water on its surface. That we know of—it's the ultimate oasis!

Mars has the tallest volcano in the solar system, Olympus Mons. It's 13 miles high—talk about a mountain with altitude!

Jupiter's Great Red Spot is a giant storm that's been raging for over 300 years. It's like a hurricane that just won't quit.

Saturn's rings are made of ice, rock, and dust. They're basically the galaxy's most stylish accessories!

Uranus rotates on its side. It's like a planet that just decided to take a nap.

Neptune has the fastest winds in the solar system, reaching up to 1,200 mph. Hold on to your hat—or your spaceship!

Pluto is smaller than Earth's moon. But it's still the fan-favorite "dwarf planet"!

Venus is the hottest planet in the solar system, with temperatures over 900°F. It's like a pizza oven in space!

Mars has seasons like Earth, but they're twice as long. Double the winter sounds fun—until you realize it's really cold!

Jupiter has at least 95 moons. It's like the popular kid at school—everyone wants to orbit around it!

Uranus appears blue-green because of methane gas in its atmosphere. Who knew gas could be so pretty?

Neptune was the first planet discovered using a telescope. It was like finding treasure in the night sky!

Earth is the densest planet in the solar system. We may be heavy, but we're proud of it!

Saturn is so light it could float in water—if you could find a tub big enough! Now that's some serious buoyancy.

Mars has two tiny moons named Phobos and Deimos. They're like little space potatoes orbiting the planet.

Jupiter's magnetic field is the strongest of all the planets. It's like the bodybuilder of the solar system!

Pluto has a heart-shaped glacier made of nitrogen ice. It's the cutest planet feature in the galaxy!

The sun takes up 99.8% of the mass in the solar system. It's the ultimate spotlight hog—and we're okay with it!

Bizarre Buildings

The Crazy House in Vietnam twists and turns like a dream brought to life. It's less of a building and more of a real-life fairy tale!

The Eiffel Tower grows taller in the summer because metal expands in heat. Even buildings like to stretch in the sun!

Dubai's Burj Khalifa is the tallest building in the world, standing at 2,717 feet. That's one seriously tall skyscraper!

The Great Pyramid of Giza was the tallest man-made structure for over 4,000 years. It held the world record before skyscrapers came along!

The Sydney Opera House's roof was inspired by orange slices. Who knew architecture could be so delicious?

The Palace of the Parliament in Romania is the heaviest building in the world. It weighs over 4 billion kilograms—that's one heavy palace!

The Sagrada Família in Spain has been under construction for over 140 years. It's taking longer to finish than a video game download on slow Wi-Fi!

China's Bird's Nest Stadium was designed to look like, well, a bird's nest. It's cozy and cool at the same time!

The Taj Mahal was built as a tomb for a queen. It's the fanciest "I love you" gift ever!

The Colosseum in Rome could hold over 50,000 spectators. Ancient Rome really knew how to throw a party!

The Guggenheim Museum in Spain is covered in titanium. It's like a shiny spaceship landed in the middle of the city!

The Lotus Temple in India is shaped like a giant flower. Who wouldn't want to visit a building that blooms?

The Dancing House in Prague looks like it's swaying to music. It's like the building is ready to join a dance party!

The upside-down WonderWorks buildings are designed to look like they've been flipped over. They're like architectural pranks!

The Petronas Towers in Malaysia are connected by a sky bridge. It's like the ultimate walk in the clouds!

The Casa Batlló in Spain has a roof that looks like a dragon's back. It's perfect for fantasy lovers!

The Louvre Museum in Paris has a giant glass pyramid at its entrance. Art meets geometry in the fanciest way possible.

The Shard in London is the tallest building in the UK. Its pointy top looks like it could pierce the clouds!

The Vatican's St. Peter's Basilica took over 120 years to build. It's like the ultimate slow-cooked masterpiece.

The Crooked House in Poland looks like it melted. It's straight out of a fairytale—or maybe a funhouse!

Chapter 5

UNSOLVED AND UNEXPECTED

Get ready to uncover the unexpected and dive into the unexplained! From the mind-boggling mysteries of the Bermuda Triangle and ancient statues on Easter Island to the sneaky secrets of history's most famous spies, this chapter is packed with surprises. You'll learn about peculiar places, powerful forces like earthquakes and volcanoes, and dazzling crystals that sparkle with beauty and science. Packed with wonders and riddles, this chapter will leave you marveling at the mysteries that surround us every day!

Mind-Boggling Mysteries

The Bermuda Triangle is an area where ships and planes mysteriously disappear. It's like nature's ultimate "lost and found"!

The Tomb of China's first emperor, Qin Shi Huang, is guarded by thousands of life-sized Terracotta Warriors. But the emperor's actual tomb remains sealed—what secrets lie inside?

The Loch Ness Monster has been rumored to live in a Scottish lake for centuries. Is it a dinosaur, a big fish, or just a myth?

The Voynich Manuscript is a book filled with mysterious symbols no one can decode. It's the ultimate puzzle book!

Some people believe the Pyramids of Egypt were built by aliens. If that's true, those aliens were great architects!

Nobody knows what happened to the lost city of Atlantis. Was it real or just a story? Either way, it's a legend that won't sink.

The disappearance of Amelia Earhart remains unsolved. The famous pilot vanished without a trace—an aviation mystery for the ages!

Bigfoot is a legendary creature said to roam forests. Is it an ape, a prank, or just camera tricks?

The Nazca Lines in Peru are massive drawings in the desert, visible only from the sky. Who made them, and why?

The mystery of Oak Island revolves around a hidden treasure no one has found yet. It's like a real-life pirate map!

The Tunguska event was an explosion in Siberia in 1908, and no one knows what caused it. A meteor? A UFO? Nature's firework?

The Dancing Plague of 1518 caused people to dance uncontrollably for days. Was it stress, a sickness, or just a really catchy tune?

The identity of Jack the Ripper has never been discovered. This infamous criminal left behind a trail of mystery.

The Wow! Signal was a strange radio signal from space in 1977. Was it aliens saying hello?

Easter Island statues were moved into place somehow—but how? Rolling logs? Alien cranes? Nobody knows for sure!

The Mary Celeste ship was found abandoned in perfect condition. The crew was gone, but their coffee was still warm!

Crop circles appear mysteriously in fields. Are they pranks or messages from outer space?

The mystery of Roanoke involves an entire colony that vanished. All that was left was a cryptic word carved into a tree.

The Zodiac Killer sent coded messages that still haven't all been cracked. It's a puzzle that remains unsolved.

The lost Amber Room, a room made of amber and gold, disappeared during World War II. It's one of history's shiniest mysteries!

Sneaky Spies

The CIA once created a spy cat named "Acoustic Kitty" to eavesdrop on conversations. Imagine being followed by a secret agent with whiskers!

Mata Hari was a famous spy during World War I who used her charm to gather secrets. She made espionage glamorous!

The Enigma machine was used by Germany during World War II to send coded messages. Breaking the code helped win the war!

James Bond is based on real-life spies from the Cold War. The gadgets, however, might be a little exaggerated!

Spies often used hollowed-out coins to hide messages. Talk about pocket change with a purpose!

The Soviet Union built tiny cameras to hide in pens and buttons. Everyday objects turned into secret spy gear!

Spy pigeons were trained to carry messages during World War I. They were feathered heroes!

The "Cambridge Five" were a group of British spies who worked for the Soviet Union. Double agents with double lives!

Some spies used invisible ink made from lemon juice. Heat it up, and the message appears—like magic!

In World War II, Monopoly games were used to smuggle escape tools to prisoners. Freeing people, one board game at a time!

Spy satellites can see objects as small as a car from space. It's like Google Earth, but super sneaky!

Nazi spies once tried to sabotage American factories during World War II. Their plans were foiled by FBI agents.

Cryptography, or codebreaking, is one of the most important spy skills. It's like solving puzzles to save the world!

Spies often wore disguises, including fake mustaches and wigs. Espionage meets costume party!

The KGB, the Soviet Union's spy agency, was one of the most feared in history. They took sneaky to a whole new level.

During the Revolutionary War, women acted as spies by sewing messages into their clothing. Secret stitching for the win!

The CIA once made fake dragonflies as tiny drones. Nature-inspired espionage!

Some spies use microdots—tiny photographs that are almost impossible to see. The ultimate tiny secret!

Spy planes like the U-2 can fly at extreme altitudes to gather intelligence. High above and undercover!

Espionage has been around for thousands of years, even in ancient Egypt. Spies are the original secret keepers!

Quirky Quakes

The largest earthquake ever recorded was a magnitude 9.5 in Chile in 1960. That's enough to shake your socks off—literally!

Earthquakes are caused by the movement of tectonic plates. It's like Earth's version of rearranging the furniture.

The "Ring of Fire" is an area with lots of earthquakes and volcanoes. Sounds dangerous, but it's also pretty cool!

Volcanoes erupt when magma builds up pressure underground. It's like a soda bottle ready to explode!

Mount St. Helens erupted in 1980, blowing off its entire top. Talk about a mountain makeover!

Tsunamis are giant waves caused by underwater earthquakes. They can travel faster than a jet plane!

The San Andreas Fault in California is one of the most famous fault lines. It's a crack in the Earth you don't want to mess with!

Some earthquakes are so small that only special instruments can detect them. Tiny quakes are Earth's quiet rumblings!

The lava from Kilauea volcano in Hawaii can flow for miles. It's like nature's slow-moving fire river.

Volcanic ash can travel thousands of miles in the air. It's like the Earth sneezed and spread ash everywhere!

Hot springs are heated by underground volcanic activity. Nature's hot tubs are always open!

Some volcanic eruptions create lightning in the ash cloud. It's like a firework show in the sky.

Earthquakes can cause the ground to liquefy. Solid ground turning to mush? Spooky!

Volcanoes can form islands, like the Hawaiian Islands. Lava works slow, but it builds big!

The Pacific Ocean has the most volcanoes of any ocean. It's the ultimate volcano hotspot!

Volcanic eruptions can create colorful sunsets. Beauty after the blast!

The Yellowstone supervolcano could erupt again someday. Let's hope it keeps taking a long nap!

Some animals can sense earthquakes before they happen. Nature's furry and feathery warning system!

Volcanic glass, like obsidian, forms when lava cools quickly. It's shiny, sharp, and super cool!

Some lava flows so fast that it can outrun a sprinting human. Once it starts moving, there's no stopping that river of fire!

Crazy Crystals

Diamonds are made from carbon under extreme pressure. So basically, they're just fancy, shiny rocks!

Quartz is one of the most common crystals on Earth. It's everywhere—like nature's glitter!

Amethyst is a purple quartz once believed to prevent drunkenness. A crystal cure for over-partying!

Salt is a crystal, and you probably eat it every day. Who knew your fries were seasoned with science?

Snowflakes are tiny ice crystals, and no two are alike. Every snowflake is like nature's fingerprint!

Some crystals grow underwater in caves, like giant gypsum crystals. They're like secret treasures hidden deep underground.

Geodes are hollow rocks lined with crystals inside. They're like nature's surprise eggs!

Emeralds are green because of the element chromium. A little bit of chemistry creates a lot of sparkle!

Crystals form in a process called crystallization when liquids cool and harden. It's like a magic freeze!

Rubies and sapphires are both made of the same mineral: corundum. One's red, the other's blue—it's like choosing a team!

Obsidian is volcanic glass formed when lava cools quickly. It's sharp enough to be used as ancient tools!

Crystals can grow larger when there's more time and space. Patience pays off in the crystal world!

The Hope Diamond is one of the most famous gems in the world. It's said to be beautiful—and cursed!

Crystal caves in Mexico hold some of the largest crystals ever found. They're taller than a house!

Topaz comes in many colors, but the blue variety is the most popular. It's like having a rainbow of choices.

Crystal structures are so precise that scientists study them to learn about math and symmetry. Nature is the ultimate artist and mathematician!

Opals are made of tiny spheres that reflect light to create a rainbow effect. They're like mini kaleidoscopes!

Some crystals glow under UV light, like fluorite. It's like having your own glow-in-the-dark treasure!

Turquoise has been used in jewelry for thousands of years. It's the ancient world's favorite pop of color.

Crystals are used in electronics, like quartz in watches. Science keeps time sparklingly accurate!

Stunning Statues

The Statue of Liberty was shipped from France to the U.S. in pieces. It was the world's fanciest puzzle!

The Colossus of Rhodes was a giant statue of the Greek sun god Helios. It stood tall—until an earthquake took it down.

The Christ the Redeemer statue in Brazil is 98 feet tall. That's almost as tall as a 10-story building!

Mount Rushmore features 4 U.S. Presidents carved into a mountain. It's like a giant historical selfie!

The Moai statues on Easter Island are so heavy they had to be "walked" into place. How did they do it? It's still a mystery!

Michelangelo carved the famous statue of David from a single block of marble. One rock, one masterpiece!

The Great Sphinx of Giza has been guarding the pyramids for thousands of years. It's the ultimate watchdog!

The Venus de Milo is missing her arms, but she's still a work of art. Even statues have bad days!

The Terracotta Army in China has over 8,000 life-sized statues of soldiers. That's a lot of ancient teamwork!

The Little Mermaid statue in Denmark is based on the fairy tale. She's much smaller than most statues—only 4 feet tall!

The Thinker by Auguste Rodin is one of the most famous statues of all time. He looks like he's trying to solve a really hard math problem!

The Statue of Unity in India is the tallest statue in the world, at 597 feet. It's like stacking five Statues of Liberty!

The Winged Victory of Samothrace is a statue of a headless goddess. She's still stunning, even without a head!

The ancient Greeks often painted their statues in bright colors. White marble was never their final look—talk about colorful creativity!

The Angel of the North in England is made of 200 tons of steel. It's an angel built to last!

The Easter Island statues have bodies buried underground. Surprise—they're not just heads!

The Reclining Buddha statue in Thailand is over 150 feet long. That's one relaxed statue!

The Statue of Zeus at Olympia was one of the Seven Wonders of the Ancient World. It was made of gold and ivory—talk about luxury!

The Pietà by Michelangelo shows Mary holding Jesus. It's considered one of the most emotional statues ever carved.

The Crazy Horse Memorial in the U.S. is still being carved. When finished, it will be bigger than Mount Rushmore!

Chapter 6

THE POWER OF HUMAN INGENUITY

Humans are endlessly creative, building bridges that defy gravity, designing trains that zoom faster than the wind, and dreaming up inventions that change the world. In this chapter, you'll explore incredible innovations, mind-blowing technology, and even some of the strangest jobs out there. Discover how engineers craft towering bridges, how gadgets like smartphones came to be, and why trains are more remarkable than you ever imagined. This chapter celebrates the power of human ingenuity and the wild, wonderful ideas that keep our world moving forward!

Incredible Inventions

The first lightbulb was invented by Thomas Edison in 1879. Before that, it was candles and lanterns all the way!

The airplane was invented by the Wright brothers in 1903. Imagine inventing something that could actually fly!

The telegraph, invented by Samuel Morse in the 1830s, revolutionized long-distance communication. It was the first "instant message"!

The first computer, ENIAC, was as big as a room. Now, we carry more power in our pockets!

The microwave oven was accidentally invented when a scientist noticed a chocolate bar melted in his pocket. A sweet discovery, literally!

Velcro was inspired by burrs that stuck to a man's clothes. Nature's little hitchhikers became a handy invention!

The first mechanical clock was invented in the 14th century. Imagine always being on time for the first time in history!

The invention of paper originated in China over 2,000 years ago. Life without notebooks? Unthinkable!

The telephone was invented by Alexander Graham Bell in 1876. Before that, people had to write letters—or shout really loud!

The first video game, "Pong," was created in 1972. It was just a bouncing dot, but it was revolutionary!

The wheel was invented around 3500 BCE. It's still rolling strong today!

The invention of penicillin by Alexander Fleming in 1928 changed medicine forever. A little mold saved millions of lives!

The internet was invented in the 1960s to help scientists share information. Now it's a place for memes, too!

The zipper was invented in 1913. It's the unsung hero of clothing!

The first camera was so big it needed its own room. And taking one photo took hours—say cheese for a long time!

The first 3D printer was invented in 1984. Now you can print almost anything—even pizza!

Bubble wrap was originally invented as wallpaper. Turns out, popping it is way more fun!

The bicycle was invented in the early 1800s. The original eco-friendly ride!

Windshield wipers were invented by a woman, Mary Anderson, in 1903. She made rainy drives much safer!

The first robot was invented in 1495 by Leonardo da Vinci. It wasn't R2-D2, but it was a start!

Terrific Technology

Smartphones are more powerful than the computers that sent astronauts to the moon. Imagine texting with that kind of tech!

The first email was sent in 1971. It was probably the world's first "You've got mail!" moment.

The GPS system relies on at least 24 satellites orbiting the Earth. Without it, road trips would still require paper maps!

Drones are used for deliveries, photography, and even saving lives. Flying robots are here to stay!

Bluetooth was named after a Viking king. Who knew technology could have historical roots?

The first video call happened in 1964 at the World's Fair. It was like FaceTime's great-grandparent!

Wi-Fi stands for "Wireless Fidelity." It's the reason we can connect to the internet without cables!

Robots are used to perform surgeries with extreme precision. They're the ultimate steady hands!

Virtual reality lets you explore worlds without leaving your room. It's like teleportation, but with goggles.

The first ATM was introduced in 1967. Imagine a world where cash wasn't so easy to grab!

Self-driving cars use cameras and sensors to navigate. The future is here, and it's on autopilot!

The first smartwatch was invented in the 1970s. It could tell time and solve math problems—how handy!

Artificial intelligence can write songs, paint pictures, and even play chess. But can it make better jokes than humans?

3D glasses work by showing each eye a slightly different image. It's like tricking your brain into seeing depth!

Electric cars were invented in the 1800s but became popular again in the 21st century. They're old-school tech with a modern twist!

A robotic vacuum cleaner was first sold in 2002. Finally, robots that help with chores!

Nanotechnology deals with things so tiny they're measured in billionths of a meter. It's science on a super small scale!

QR codes were first used to track car parts. Now they're everywhere—even on menus!

The International Space Station is the largest man-made object in space. It's like a giant science lab floating above Earth!

Smart homes can control lights, locks, and even coffee makers with voice commands. It's the ultimate life hack!

Jazzy Jobs

Ice cream tasters test flavors all day. Imagine having that as your job—dreamy!

Snake milkers collect venom to make life-saving medicine. Dangerous, but heroic!

Professional cuddlers offer hugs to people who need comfort. It's a job built on kindness!

Dog surfers train pups to ride waves in competitions. The coolest job on the beach!

Cricket farmers raise insects for food. It's a job that's hopping with possibilities.

Toy testers make sure toys are fun and safe. Best job for kids-at-heart!

Water slide testers check speed and safety. Splashing into work sounds amazing!

Space archaeologists study artifacts in outer space. History meets sci-fi!

Professional whistlers perform in movies and music. Who knew whistling could be a career?

Odor testers sniff things like deodorants to ensure they work. A job that stinks—literally!

Penguin caretakers feed and study penguins in Antarctica. A chilly but adorable career!

Video game designers create worlds for players to explore. Turning imagination into reality!

Foley artists create sound effects for movies. They make footsteps, punches, and even dinosaur roars!

Shark trackers study shark movements to protect them and their habitats. Working with the ocean's top predators!

Theme park designers build roller coasters and attractions. The masterminds behind the thrills!

Food stylists make meals look picture-perfect for ads. Their secret weapon? Glue instead of milk!

Bee farmers raise bees for honey and pollination. Nature's tiny helpers need human friends!

Chocolate tasters sample sweets for quality control. Talk about a sugar rush at work!

Face painters make art on people's cheeks at festivals. It's art that smiles back!

Professional mermaids perform underwater shows. It's like living in a fairy tale!

Remarkable Railways

The Trans-Siberian Railway is the longest in the world, stretching over 5,700 miles. It's like traveling across an entire continent by train!

Japan's bullet trains, or Shinkansen, can travel up to 200 mph. Talk about a speedy commute!

The Hogwarts Express in the Harry Potter movies was filmed on a real train in Scotland. Magic and railways make the perfect combo!

The first steam-powered train was invented in 1804. It was slow, but it was the start of something big!

The Channel Tunnel connects England and France underwater. It's a train ride through the sea!

Luxury trains like the Orient Express offer fine dining and fancy cabins. It's travel that feels like a 5-star hotel.

The Glacier Express in Switzerland travels through stunning mountain landscapes. It's called "the world's slowest express train" because the views are worth savoring!

Some trains are powered by magnets, like maglev trains. They don't even touch the tracks—they hover!

The Indian Railways system is one of the largest employers in the world. Running trains takes a LOT of team-work!

The Darjeeling Himalayan Railway is so steep that the train zigzags to climb the mountains. It's like a roller coaster on tracks!

The world's highest railway bridge is in India, taller than the Eiffel Tower. Who said bridges can't be bold?

Train tracks can stretch across deserts, jungles, and even snowy tundras. There's no terrain they can't conquer!

Subway systems in cities like New York and Tokyo carry millions of passengers daily. They're underground marvels!

The first electric train was built in 1879 in Berlin. It's the eco-friendly grandparent of modern railways!

Trains in the Netherlands run entirely on wind power. They're literally powered by breezy days!

The fastest train in the world, Shanghai's maglev, can reach 267 mph. That's faster than most airplanes during takeoff!

Train horns are tuned to specific musical notes. It's like trains are singing their way down the tracks!

Some trains have glass ceilings for passengers to enjoy panoramic views. Sightseeing has never been so comfy!

Australia's Indian Pacific train crosses the longest straight stretch of track in the world—297 miles without a single curve! No twists, no turns, just straight ahead.

The world's smallest train station is in Japan and serves just one passenger. Now that's personal service!

Bold Bridges

The Golden Gate Bridge in San Francisco is painted "international orange." It's not gold, but it sure is iconic!

The Millau Viaduct in France is the tallest bridge in the world. It's so tall, it sometimes rises above the clouds!

The Brooklyn Bridge was completed in 1883 and was the first bridge to use steel cables. It's an engineering pioneer!

The Tower Bridge in London has two towers connected by walkways. It's like a castle and a bridge combined!

The Akashi Kaikyo Bridge in Japan is the longest suspension bridge in the world. It's over 12,000 feet long—talk about a long walk!

The Sydney Harbour Bridge is nicknamed "The Coathanger" because of its shape. It's stylish and functional!

The Bixby Creek Bridge in California has been featured in many movies and TV shows. It's Hollywood's favorite bridge!

The Confederation Bridge in Canada is the longest bridge over ice-covered waters. A chilly but impressive connection!

The Banpo Bridge in South Korea has a fountain that lights up at night. It's part bridge, part water show!

The Iron Bridge in England was the first bridge made of cast iron. It's been standing strong since 1779!

The Helix Bridge in Singapore is shaped like a DNA strand. It's science and engineering rolled into one!

The Charles Bridge in Prague is lined with 30 statues of saints. Crossing it feels like walking through history!

The Mackinac Bridge in Michigan connects two peninsulas. It's nicknamed "Big Mac," but it's not about burgers!

The Øresund Bridge connects Denmark and Sweden. It's part bridge, part tunnel—it's a two-for-one marvel!

The Rainbow Bridge in Tokyo lights up in a rainbow of colors at night. It's a bridge and a light show!

The Rialto Bridge in Venice is over 400 years old. It's one of the city's most picturesque landmarks!

The Bridge of Sighs in Venice got its name because prisoners sighed as they crossed it on their way to jail. Romantic, but with a dramatic twist!

The Zhangjiajie Glass Bridge in China is transparent. Walking on it feels like stepping into the sky!

The Forth Bridge in Scotland is a UNESCO World Heritage Site. It's an iron masterpiece over 130 years old!

The Eshima Ohashi Bridge in Japan looks like a roller coaster. It's so steep, it's both thrilling and functional!

Chapter 7

PLAYFUL PERSPECTIVES: PETS, GAMES, AND MORE!

This chapter is packed with playful and curious surprises, from adorable pets and record-breaking feats to mind-blowing games and the magic of music. Discover quirky jobs, fascinating word origins, and how our love of creativity and fun connects us across cultures and history. Whether it's the wag of a dog's tail, the beauty of a musical instrument, or the laughter sparked by a witty word, these facts celebrate the lighter side of life. Prepare to be amazed, inspired, and thoroughly entertained!

Playful Pets

A dog's nose print is as unique as a human fingerprint. It's like their own little ID card!

Cats can rotate their ears 180 degrees. That's some serious eavesdropping power!

Goldfish can remember things for months, not just seconds. So much for the "forgetful fish" myth!

Hamsters run up to 5 miles a night on their wheels. Who needs a gym when you're a hamster?

Parrots can learn to mimic human speech. They're nature's chatty comedians!

Dogs can be trained to detect diseases like cancer. They're not just cute—they're lifesavers!

Rabbits' teeth never stop growing. That's why they love to chew—it's dental care for them!

Guinea pigs "popcorn" when they're happy. It's like they're jumping for joy—literally!

Cats spend about 70% of their lives sleeping. Living the ultimate lazy life!

Some turtles can live to be over 100 years old. Imagine celebrating a century of birthdays!

Dogs can understand up to 250 words and gestures. That's more than some toddlers!

Ferrets love to hide things in small spaces. They're tiny, adorable hoarders!

Horses can communicate using facial expressions. They're like four-legged emojis!

Budgies, a type of parakeet, are among the best talkers in the bird world. Tiny birds with big vocabularies!

Pet pigs are incredibly smart and can even play video games. Who knew pigs could be gamers?

Dogs wag their tails to the right when they're happy. Tail wagging is like their version of smiling!

Some cats are naturally left-pawed. Just like humans, they can be lefties or righties!

Axolotls, often kept as pets, can regrow their limbs. They're like the superheroes of the pet world!

Birds like cockatiels can recognize themselves in mirrors. They're little feathered narcissists!

Chinchillas have the softest fur in the animal kingdom. It's so dense, they can't even get fleas!

Fantastic Feats

The fastest human can run 27.5 mph. That's faster than most people drive in a school zone!

The longest anyone has held their breath underwater is over 24 minutes. That's like an episode of your favorite show without breathing!

The tallest man ever recorded was 8 feet 11 inches tall. He had to duck through every doorway!

The smallest waist ever recorded was just 13 inches. That's smaller than a basketball hoop!

The longest fingernails ever grew to over 28 feet combined. Imagine trying to text with those!

The longest time spent balancing on one foot is over 76 hours. Talk about dedication—and sore legs!

The heaviest weight ever lifted by a human is over 6,000 pounds. That's like lifting a small car!

The world record for eating hot dogs is 76 in 10 minutes. Don't try this at home!

The longest hair ever recorded measured over 18 feet. That's a lot of shampoo and conditioner!

The youngest Nobel Prize winner was just 17 years old. I'm sure I wasn't doing that at 17!

The farthest a paper airplane has flown is 226 feet. That's longer than a football field!

The highest anyone has jumped without equipment is over 8 feet. That's like hopping over a giraffe's head!

The largest pizza ever made was over 13,000 square feet. That's enough to feed an entire city!

The fastest talker can say over 600 words per minute. Good luck understanding that!

The longest time someone has spent awake is 11 days. Hopefully, they had a great nap afterward!

The longest someone has juggled is over 12 hours. That's juggling through an entire day!

The most push-ups done in 24 hours is over 46,000. My arms hurt just thinking about it!

The world's largest sandcastle was over 57 feet tall. That's a castle fit for a sandy king!

The longest hiccuping spree lasted for 68 years. Imagine trying to hold your breath for that long!

The largest rubber band ball weighed over 9,000 pounds. That's one bouncy creation!

Glorious Games

Chess is over 1,500 years old. It's been checkmating brains for centuries!

The longest Monopoly game lasted 70 days. That's more rent than anyone wants to pay!

Jenga was invented using blocks from a woodpile in Africa. A simple game with wobbly stakes!

Tag is one of the oldest games in history, dating back to ancient Greece. You're never too old for a game of tag!

The first video game, "Pong," was just a bouncing dot. It started a gaming revolution!

Tug-of-war used to be an Olympic sport. Imagine winning a gold medal for pulling a rope!

Rubik's Cubes have over 43 quintillion possible combinations. Solving one is basically magic!

Hide and seek is played all over the world. It's the universal game of disappearing!

Checkers dates back to 3,000 BCE in ancient Mesopotamia. It's been hopping along for thousands of years!

The first soccer ball was made of inflated pig bladders. Ew, but also kind of genius!

Video games are played by over 3 billion people worldwide. That's almost half the planet!

Dominoes originated in China around the 12th century. Now they're toppling around the globe!

The first playing cards were invented in China in the 9th century. Card games go way back!

Basketball was invented by a gym teacher using a peach basket. Talk about humble beginnings!

Scrabble players memorize thousands of two-letter words to win. Who knew "qi" would be so handy?

The first board game, "Senet," was played in ancient Egypt. Even the pharaohs liked game night!

Rock-paper-scissors originated in Japan. It's the greatest decision-maker!

The tallest stack of dice in the world is 20 high. Don't sneeze near that!

Pinball was banned in the U.S. for 30 years because it was considered gambling. Who knew pinball was so controversial?

Pac-Man was originally called "Puck Man" in Japan. They changed it to avoid a certain... mispronunciation!

Musical Magic

The world's largest piano is over 18 feet long. Now that's a truly grand piano!

The didgeridoo is one of the world's oldest instruments, invented by Indigenous Australians. It makes a sound like no other!

Violins were originally called fiddles. Fiddling around turned into high art!

The theremin is an instrument you play without touching it. It's like music made by magic!

Bagpipes originated in ancient Egypt. They've been piping up for thousands of years!

The piano has over 12,000 parts. That's a lot of keys, strings, and magic!

The largest drum in the world weighs over 7 tons. That's a beat you can really feel!

Hummingbirds can hum because their wings flap so fast. Nature's tiniest musicians!

The oldest known flute is over 40,000 years old. Music has been part of life since the Stone Age!

A guitar's strings were originally made of animal guts. Thankfully, they're made of metal or nylon now!

The world's largest orchestra had over 8,000 musicians. That's a symphony of epic proportions!

Drums are the oldest musical instrument. Rhythm has always been in our DNA!

The glass harmonica makes music by rubbing your fingers on spinning glass bowls. It's like making wine glasses sing!

The kazoo was invented in the 1840s. It's the silliest instrument with the funniest sound!

The marimba is a giant xylophone from Central America. It's big, bold, and beautiful!

Tuning forks vibrate at a specific pitch. They're like little metal music makers!

Accordion players press buttons and pull to create sound. It's like a one-person band!

Some trumpets are made of vibrant plastic but still sound amazing. Who says music can't be colorful?

Orchestras often tune to an oboe because its pitch is steady. The oboe is the boss of the orchestra!

Harps can have up to 47 strings. That's a lot of plucking power!

Witty Words

The longest word in English is 189,819 letters long. It's the name of a protein—good luck pronouncing it!

The word "alphabet" comes from the first two Greek letters: alpha and beta. It's the original A-B-Cs!

"Emoji" is a Japanese word that means "picture character." Emojis are basically tiny word art!

The word "quiz" was supposedly invented as part of a bet. Imagine making up a word just for fun!

Palindromes are words or phrases that read the same backward and forward, like "racecar." They're the word version of a boomerang!

The most common letter in English is "E." It's everywhere—you just can't escape it!

The word "goodbye" comes from the phrase "God be with ye." Even old-timey goodbyes were polite!

The word "kangaroo" comes from an Aboriginal Australian word meaning "I don't know." It's a big misunderstanding that stuck!

A "pangram" is a sentence that uses every letter of the alphabet, like "The quick brown fox jumps over the lazy dog." It's the alphabet's greatest hits!

The dot over the letters "i" and "j" is called a "tittle." Even dots have names!

The English language adds a new word every two hours. Dictionaries have a full-time job keeping up!

The word "OK" is one of the most recognized words in the world. It's short, simple, and... OK!

There's a word for the fear of long words: "hippopotomonstrosesquipedaliophobia." It's ironically long and terrifying!

The word "sandwich" is named after the Earl of Sandwich. He wanted to eat without leaving the poker table—priorities!

"Spoonerisms" are when you mix up the first sounds of two words, like "runny babbit" instead of "bunny rabbit." It's wordplay that's wonderfully wacky!

The word "gymnasium" comes from a Greek word meaning "to exercise naked." Glad we've updated that rule!

"Oxymoron" means putting two opposite words together, like "jumbo shrimp." It's wordplay with a twist!

English is the only language where "pineapple" doesn't sound like "ananas." We had to be different!

There's a word for mishearing lyrics: "mondegreen." Like singing, "Hold me closer, Tony Danza" instead of "Tiny Dancer"!

The word "ghost" is from an old German word that means "spirit." It's spooky no matter the language!

Chapter 8

FROM STARS TO STATUES: WONDERS OF THE UNIVERSE

From the depths of space to the wonders of Earth, this chapter explores the universe's most captivating mysteries and marvels. Discover spectacular stars, peculiar planets, and the vibrant cultures that make life on Earth extraordinary. Learn about quirky vehicles that defy imagination and legendary landmarks that tell stories of our world's history. Whether it's the grandeur of the cosmos or the charm of human creativity, this chapter celebrates the awe-inspiring wonders of the universe!

Spectacular Space

A day on Venus is longer than a year on Venus. Imagine celebrating your birthday for an entire year!

There are more stars in the universe than grains of sand on all Earth's beaches. That's a lot of cosmic grains!

The Milky Way galaxy is so big it would take 100,000 years to travel across it at the speed of light. Better pack a LOT of snacks for that trip!

Black holes have such strong gravity that they can warp time itself. They're like cosmic trapdoors to the unknown!

Neutron stars are so dense, a sugar-cube-sized piece would weigh as much as a mountain. That's one heavy cube!

The moon is slowly drifting away from Earth at about 1.5 inches per year. Don't worry—it'll take billions of years to really get far!

There's a giant storm on Jupiter that's been raging for over 300 years. Talk about extreme weather!

The sun's core is 27 million degrees Fahrenheit. That's definitely hotter than your summer BBQ!

There's a planet made entirely of diamonds called 55 Cancri e. The ultimate sparkling mystery!

Saturn's moon Titan has lakes of liquid methane instead of water. Imagine swimming in that—it'd be slippery!

A year on Neptune lasts 165 Earth years. You'd be waiting a LONG time for your next birthday.

The largest volcano in the solar system, Olympus Mons, is on Mars. It's three times taller than Mount Everest!

The temperature on Mercury can swing from 800°F during the day to -290°F at night. Pack for all seasons!

Stars don't twinkle—they only look like they do because of Earth's atmosphere. Space is less sparkly than it seems!

The Great Attractor is a mysterious force pulling galaxies toward it. It's the ultimate cosmic magnet.

There's a "cold spot" in space that scientists can't explain. It's like the universe's version of a chilly mystery!

A comet's tail always points away from the sun. It's like space's windblown hairstyle!

The sun takes 225 million years to orbit the center of the Milky Way. It's on a very slow space road trip!

Some stars explode into supernovas, creating elements like gold and silver. Thanks, stars, for our shiny stuff!

The moon doesn't have wind or water, so astronauts' footprints can last for millions of years. Talk about leaving a lasting impression!

Peculiar Planets

Mercury doesn't have an atmosphere, so its skies are pitch black. Even during the day, it's all-night vibes!

Uranus is the coldest planet in our solar system. It's like a frozen popsicle in space.

Mars has the largest dust storms in the solar system. You'd definitely need goggles there!

Venus is covered in thick clouds of sulfuric acid. A literal acid rain nightmare!

Jupiter has a moon, Ganymede, that's larger than Mercury. A moon bigger than a planet—mind blown!

Saturn's rings are made of billions of ice particles. It's like the galaxy's biggest snow globe!

Pluto's heart-shaped glacier is made of nitrogen ice. The most adorable feature in the universe!

Earth is the only planet with tectonic plates that move. We're the movers and shakers of the solar system!

Neptune has faint rings made of dust. Even planets love accessorizing!

There are "hot Jupiters," gas giants outside our solar system that orbit their stars super close. Think Jupiter, but sizzling!

Mercury doesn't have seasons because it barely tilts on its axis. It's the solar system's most predictable planet!

Venus is the hottest planet in the solar system, with temperatures high enough to melt lead. Talk about a planet with a serious heatwave!

Mars has polar ice caps made of dry ice (frozen carbon dioxide). It's like a giant space freezer!

Saturn is the least dense planet—it could float in water. Too bad there's no bathtub big enough!

Jupiter's moon Europa might have a hidden ocean under its icy surface. Could there be life lurking beneath?

Uranus rotates on its side, making its seasons extreme. Imagine decades of darkness, then decades of sunlight!

Earth's magnetic field protects us from harmful solar radiation. It's like our planet's invisible shield!

Venus is the hottest planet, even though Mercury is closer to the sun. Blame Venus's thick atmosphere for the heat!

There are over 5,000 confirmed exoplanets outside our solar system. The universe is teeming with possibilities!

Neptune has winds that blow over 1,000 mph. Hold onto your space helmets!

Colorful Cultures

In Japan, slurping noodles is a sign of appreciation. It's polite AND deliciously noisy!

In India, throwing colorful powders during Holi celebrates joy and love. It's a rainbow party for everyone!

In Spain, La Tomatina is a festival where people throw tomatoes at each other. It's the world's squishiest food fight!

In Iceland, people give books as gifts on Christmas Eve and spend the night reading. Cozy AND thoughtful!

In Scotland, tossing a tree trunk (called a caber) is a traditional sport. Tree lifting, anyone?

In Mexico, breaking a piñata is a birthday tradition. It's candy-filled fun with a stick!

In Brazil, the Carnival parade is full of music, dancing, and elaborate costumes. It's a celebration like no other!

In South Korea, people eat seaweed soup on their birthday. It's a meal for good luck and health!

In China, red envelopes filled with money are given during Lunar New Year. It's the happiest little red gift!

In Italy, people eat lentils on New Year's Eve for good fortune. The more you eat, the luckier you get!

In Germany, people touch chimney sweeps for good luck. Sooty, but lucky!

In Finland, there's a wife-carrying championship where the prize is beer. Teamwork at its funniest!

In Greece, smashing plates at celebrations is a way to bring good luck. Clean-up must be a mess!

In Canada, maple syrup festivals celebrate the sweet treat. Pancakes not included—bring your own!

In Turkey, drinking coffee grounds can be used to tell fortunes. Your next cup of coffee might hold your future!

In Sweden, people celebrate "Fika," a coffee-and-cake break. It's like mandatory snack time for grown-ups!

In Nigeria, masquerades are colorful and spiritual performances. Dancing masks bring traditions to life!

In the Netherlands, people ice skate on frozen canals in winter. It's the ultimate frosty playground!

In Russia, nesting dolls represent family and tradition. One doll, many surprises!

In Australia, people celebrate Christmas with barbecues on the beach. Santa in shorts sounds fun!

Vibrant Vehicles

The fastest car in the world can reach speeds over 300 mph. That's like driving faster than some planes take off!

The first bicycles were called "velocipedes" and didn't have pedals. Riders had to push themselves along with their feet—Flintstones-style!

Hot air balloons were the first form of human flight. They turn the sky into a canvas of drifting colors!

The world's longest limousine is 100 feet long and has a swimming pool inside. Talk about riding in style!

Trains in Japan can levitate using magnetic forces. Floating trains? The future is now!

The largest ship ever built is over 1,500 feet long. That's like four football fields end-to-end!

The first electric car was invented in 1828. It was ahead of its time—literally!

The space shuttle could reach speeds of 17,500 mph. That's one speedy ride to the stars!

The longest bus in the world can hold over 300 passengers. It's like a bus AND a parade float!

The smallest car ever made is only 4 feet long. Parking must be a breeze!

Monster trucks can crush cars with tires over 10 feet tall. They're the giants of the road!

Hovercrafts can glide over water and land. It's like a boat and a car combined!

The first submarines were powered by hand cranks. Imagine pedaling underwater—talk about a workout!

There are trains in Switzerland that climb mountains. It's the most scenic train ride ever!

The first airplanes didn't have seats. Passengers sat on the floor—hopefully, with seatbelts!

Luxury yachts often come with their own helicopters. Boats AND flying rides? Fancy!

The Batmobile from the movies actually works as a real car. Batman could take it for a grocery run!

Firetrucks are red because it's the easiest color to see. Plus, it looks cool in action!

Some bicycles are made entirely of bamboo. Eco-friendly AND stylish!

The largest motorcycle in the world is over 16 feet tall. Definitely not something you'd want to ride to work!

Legendary Landmarks

The Eiffel Tower was once painted yellow. It's had a few fashion changes over the years!

The Great Wall of China is so long it could wrap around the Earth at the equator. Now that's a lot of bricks!

The Taj Mahal changes color depending on the time of day. It's like nature's light show on marble!

Mount Rushmore was supposed to include the presidents' bodies, not just their faces. Imagine how much bigger it could've been!

The Leaning Tower of Pisa leans because of unstable ground. What started as a mistake became an icon!

Machu Picchu was built high in the Andes by the Inca and remained hidden for centuries. It's the ultimate lost city in the clouds!

The Parthenon in Greece has stood for over 2,000 years, surviving wars, earthquakes, and even explosions. Talk about a temple built to last!

The Pyramids of Giza are aligned with the stars. Ancient architects knew how to aim for the heavens!

The Gherkin in London isn't just a quirky name—it's designed to be energy-efficient with a shape that reduces wind resistance. A skyscraper that's as smart as it is stylish!

Stonehenge has stood for over 4,000 years, and no one knows exactly why it was built. A mystery made of rocks!

The Colosseum in Rome could flood the arena for naval battles. Gladiators AND boats—what a combo!

The Statue of Liberty's torch was originally made of gold. A shining beacon for all to see!

The Great Barrier Reef is so large it can be seen from space. It's nature's biggest underwater masterpiece!

The Hanging Temple in China is built into the side of a cliff and has stood for over 1,500 years. A temple that seems to defy gravity!

The Alhambra in Spain has over 10,000 intricate carvings. It's like a palace made of poetry.

Petra, the ancient city in Jordan, was carved entirely out of rock. It's like a giant stone puzzle!

The Forbidden City in China has over 9,000 rooms. Imagine trying to find your way in THAT palace!

Christ the Redeemer in Brazil was built on top of a mountain. The view from up there must be heavenly!

The Gateway Arch in St. Louis is the tallest arch in the world. It's a perfect 630-foot curve!

Angkor Wat in Cambodia is the largest religious monument in the world. It's like a city built for the gods!

Chapter 9

THE WILD WORLD OF WEATHER

Get ready to explore the fascinating world of weather! From wacky weather phenomena like raining fish and thundersnow to the science behind rainbows and hurricanes, this chapter dives into the mysteries of the skies. Discover the incredible power of storms, the secrets of the atmosphere, and how we predict the unpredictable. Whether it's wild winds, fabulous forecasts, or jaw-dropping lightning facts, you'll be amazed by the wonders of weather all around us!

Wacky Weather

Raining fish is a real phenomenon in some places, like Honduras. It's not just a saying—fishy weather happens!

Blood rain occurs when red dust or algae mix with raindrops. It's like the sky is painting the town red!

Ball lightning looks like glowing orbs floating through the air. It's a mystery wrapped in a stormy glow!

Ice spikes can grow out of freezing water in your freezer. Mini glaciers right in your kitchen!

Snow can be pink or red, thanks to algae. It's nicknamed "watermelon snow" and even smells sweet!

The highest temperature ever recorded was 134°F in Death Valley, California. That's hot enough to fry an egg on the sidewalk!

The coldest temperature ever recorded was -128.6°F in Antarctica. Definitely sweater weather—times 1,000!

Catatumbo lightning in Venezuela strikes the same spot almost every night. It's like nature's own strobe light!

Hailstones can grow as large as softballs. Imagine THAT crashing down on your car!

Fire tornadoes can form during wildfires. They're both terrifying and awe-inspiring.

Fog is actually a cloud that touches the ground. You're basically walking through the sky!

Dust devils are mini tornadoes made of swirling dirt. They're like little spinning dancers twirling across the landscape!

Thundersnow is when it thunders during a snowstorm. It's the ultimate weather mashup!

Waterspouts are tornadoes that form over water. They're like twirling fountains in the ocean!

Snow rollers are round balls of snow formed by wind. Nature's version of rolling snowmen!

Blue jets are lightning that shoots up into space. They're like electric rockets!

Frost quakes happen when frozen soil suddenly cracks. It's like the ground is shivering!

A heat burst is a sudden rise in temperature at night. It's like summer sneaking into your evening!

The world's windiest place is Antarctica. Bring your heaviest coat—it's a blustery adventure!

Rain can fall in the form of tiny ice pellets called graupel. It's like hail's soft-spoken cousin.

Wild Weather Wonders

Rainbows happen when sunlight passes through raindrops. It's nature's most colorful science experiment!

A double rainbow is when light reflects twice inside raindrops. Double the beauty, double the fun!

A tornado once picked up a house and set it down almost perfectly intact. Mother Nature's version of extreme home relocation!

Lightning is hotter than the surface of the sun. It's nature's ultimate spark!

A sun dog is a bright spot of light on either side of the sun. It's like the sun has its own little rainbow pets!

Auroras, or Northern Lights, are caused by charged particles from the sun hitting Earth's atmosphere. It's a cosmic light show!

Clouds can weigh millions of pounds. Imagine carrying THAT around!

Tropical cyclones are called hurricanes, typhoons, or cyclones, depending on where they occur. Same storm, different names!

A microburst is a small, intense downdraft of wind. It's like a super-strong gust from above.

Lightning can strike the same place more than once. So much for that old myth!

Snowflakes form in hexagonal shapes but are always unique. No two are alike—just like fingerprints!

Tornadoes can make objects fly through the air, including cars and even trains. Nature's strongest vacuum!

A rainbow can actually form a full circle, but we usually see just half. The rest is hiding below the horizon!

Haboobs are massive dust storms common in deserts. They're like walls of dirt on the move!

A fire rainbow isn't a rainbow—it's sunlight passing through ice crystals in clouds. It's still fiery and fabulous!

Lightning strikes Earth about 8 million times a day. That's a lot of zaps!

Tornadoes are most frequently in the U.S, especially in "Tornado Alley." It's the whirlwind capital of the world!

The greenish tint in the sky before a storm signals hail. It's nature's eerie warning!

Snow can fall even when it's above freezing if the air is dry enough. Surprise snow is the best snow!

Fog can freeze on contact with surfaces, creating "rime ice." It's like nature painting the world white.

Astounding Atmosphere

Earth's atmosphere is 78% nitrogen and 21% oxygen. The rest is a mix of fun gases like argon and carbon dioxide!

Air pressure is strongest at sea level and weaker at higher altitudes. That's why your ears pop on airplanes!

The sky is blue because of how sunlight scatters in the atmosphere. Blue just gets scattered the most!

The ozone layer protects us from the sun's harmful rays. Thanks, ozone, for being our sunscreen!

The troposphere is the lowest layer of the atmosphere, where all weather happens. It's like Earth's stage for storms!

There's no sound in space because it lacks air to carry vibrations. It's the ultimate quiet zone!

The stratosphere contains the ozone layer and jet streams. It's a busy layer up there!

Clouds form when water vapor condenses around tiny particles like dust. Even clouds need a little support!

The highest clouds, called noctilucent clouds, form near the edge of space. They're the night lights of the atmosphere!

The air gets thinner the higher you go, making it harder to breathe. That's why climbers use oxygen on Mount Everest!

The wind is caused by air moving from high-pressure areas to low-pressure areas. It's like nature evening things out.

The jet stream can travel over 200 mph and influences weather patterns. It's like an invisible river of air!

Airplanes fly above most weather in the stratosphere. Smooth skies are way up high!

Earth's atmosphere extends about 6,200 miles into space. It's a lot bigger than it looks!

The auroras happen in the thermosphere. It's the perfect layer for cosmic fireworks!

The atmosphere protects us from meteors by burning them up. Thanks for keeping us safe, air!

The greenhouse effect keeps Earth warm enough for life. It's like a planetary blanket!

Air is invisible, but it has weight—about 14.7 pounds per square inch at sea level. You're carrying a lot of air right now!

The Earth's atmosphere wasn't always full of oxygen. Early Earth was more about volcanic gases!

Satellites orbit in the exosphere, the outermost layer of the atmosphere. It's the ultimate launch pad!

Chilling Storms

Hurricanes can release as much energy as 10,000 nuclear bombs. That's one seriously powerful storm!

The largest snowflake ever recorded was 15 inches wide. Imagine that landing on your head!

A single hurricane can drop over 2.4 trillion gallons of rain. That's enough to fill millions of swimming pools!

Blizzards can cause "whiteouts," where visibility drops to almost zero. It's like being inside a snow globe!

The eye of a hurricane is calm, while the surrounding winds are the fiercest. The calm before—or during—the storm!

Typhoons are the same as hurricanes but occur in the western Pacific. Different name, same stormy power!

The Great Blizzard of 1888 dropped 55 inches of snow in some places. That's taller than a lot of people!

Tornadoes can form during hurricanes. Talk about a double weather whammy!

Hurricane winds can carry debris for miles. It's like nature's ultimate leaf blower—on steroids!

Cyclone Tracy, which hit Australia in 1974, destroyed 70% of the city of Darwin. A storm that rewrote history.

Thunderstorms produce hail when strong updrafts lift raindrops into freezing air. It's nature's icy juggling act!

The strongest hurricane winds ever recorded reached 215 mph. That's faster than most race cars!

Nor'easters are intense storms that bring snow and wind to the U.S. East Coast. They're like blizzards with an attitude!

Hurricanes spin counterclockwise in the Northern Hemisphere and clockwise in the Southern Hemisphere. A storm with a spin style!

The term "hurricane" comes from the Taino word “huracán,” meaning "god of the storm." An ancient name for a mighty force!

Snowstorms can happen even when it’s sunny, thanks to lake-effect snow. It’s like a snow machine turned on over water!

Lightning during volcanic eruptions is called “dirty lightning.” It’s a shocking addition to an eruption!

Ice storms coat everything in layers of ice, making roads and trees dangerously slippery. Beautiful, but treacherous!

Hailstones can form in layers, like onions. Each storm adds another icy ring!

The longest thunderstorm lasted over 16 hours in South Dakota. That's one long rumble in the sky!

Fabulous Forecasting

The first weather forecast was published in 1861 in a British newspaper. Before that, people just looked out the window!

Weather satellites orbit Earth to provide up-to-the-minute forecasts. They're like space-based weather reporters!

Ancient Greeks used a device called a "weather clock" to predict conditions. It's the original forecast app!

The saying "red sky at night, sailor's delight" actually has some truth to it. It's old-school weather wisdom!

Doppler radar can detect precipitation and wind. It's like the superhero of weather technology!

Weather balloons are launched daily to measure temperature, humidity, and air pressure. They're science on the rise!

The Farmer's Almanac has been predicting weather since 1818. It's like the wise, old weather grandparent!

Hurricane names alternate between male and female. And once a storm is really bad, its name gets retired forever.

Meteorologists use supercomputers to create weather models. It's weather prediction at lightning speed!

Polar orbiting satellites take detailed images of Earth's weather every 12 hours. They're like paparazzi for the atmosphere!

Frogs were once believed to predict rain. Ribbiting weather reports, anyone?

The Beaufort scale measures wind strength, from a gentle breeze to hurricane-force winds. It's a handy tool for windy days!

Snowfall predictions are tricky because snowflakes vary in size and water content. Forecasting snow is like measuring marshmallows!

Weather apps use data from thousands of stations worldwide. Your phone knows more about the weather than ever before!

Cloud types, like cirrus and cumulonimbus, can give clues about upcoming weather. Clouds are nature's weather hints!

Lightning detectors can sense strikes from hundreds of miles away. It's like eavesdropping on storms!

Weather prediction started with sailors watching the sky and sea. Early meteorologists were out at sea!

Groundhogs aren't great weather forecasters—they're right only about 40% of the time. Cute, but not reliable!

Long-range forecasts can predict trends months in advance. Science is always thinking ahead!

Weather forecasters use green screens to show maps on TV. Behind the scenes, they're pointing at nothing!

Chapter 10

SECRETS OF THE UNSEEN WORLD

Dive into the hidden and mind-blowing world of the unseen! From secret animal habitats and surprising careers to mind-bending math tricks and quantum mysteries, this chapter uncovers the incredible things you don't usually notice. Explore optical illusions that fool your brain, quantum particles that defy logic, and the fascinating science and stories hiding in plain sight. It's a journey into the invisible forces and secrets that shape our world in ways you never imagined!

Hidden Habitats

Ant colonies can have millions of ants and stretch for miles underground. It's like a tiny, bustling city!

Octopuses have three hearts and blue blood, making them one of the ocean's most unusual creatures. Talk about a real-life alien of the deep!

Tree hollows are homes to birds, squirrels, and even bees. It's like an apartment complex in the forest!

Glowworms light up dark caves to attract prey. Nature's glow sticks at work!

Coral reefs are home to 25% of all marine species. They're like the underwater version of a bustling metropolis.

Bears make dens in caves or under fallen trees to hibernate. It's the ultimate winter snooze spot!

Termites build towering mounds that can be taller than a human. It's like skyscrapers for bugs!

Arctic foxes create dens in the snow to protect their pups. Cozy and cold—true Arctic innovation!

Spiders weave webs in hidden corners to catch their meals. It's nature's version of a dinner trap!

Wombats dig burrows with multiple tunnels. They're expert architects of the animal world!

Pufferfish create intricate sand circles to attract mates. Who knew fish could be artists?

Penguins build nests out of pebbles. It's like the ultimate minimalist home!

Sloths live high up in the rainforest canopy. They're tree-dwelling experts in chilling out!

Moles dig underground tunnels to search for food. It's like a secret maze below the surface.

Bats hang out in caves during the day to rest. They're the original night shift workers of the animal world!

Clownfish live inside sea anemones for protection. It's a home that doubles as a security system!

Elephants dig waterholes in dry riverbeds. They're providing hidden hydration for everyone!

Beavers build lodges out of sticks and mud in rivers. It's like a floating fort!

Snakes often hide in holes, rocks, or under logs. They're experts at blending in!

Fiddler crabs dig burrows in the sand to escape predators. It's a beachside retreat with a view!

Secret Signals

Morse code uses dots and dashes to send messages. It's like texting… but way slower!

The first secret codes date back over 4,000 years. Ancient spies were already at work!

Carrier pigeons were used in both World Wars to deliver secret messages. Feathery little mail carriers!

The Enigma machine was a German code machine during WWII. It took a team of brilliant minds to crack it!

Some people can talk using only whistles, like the Silbo Gomero language in Spain. Imagine having an entire conversation with just whistles!

Semaphore flags use arm positions to send messages. It's like waving but with a secret meaning!

Invisible ink made from lemon juice can be revealed with heat. Spy kids, take notes!

Some pirate flags had hidden meanings, like warning of no mercy. Not just decoration—deadly serious!

Ancient Chinese warriors used drumbeats to communicate battle plans. Beats with a purpose!

Smoke signals were one of the earliest long-distance communication methods. The original text message—sent by fire!

Knights in medieval times had secret symbols on their shields. It was like wearing a team jersey in battle!

The Navajo Code Talkers used their native language to send unbreakable messages in WWII. Their code was so strong, enemies never cracked it!

Prisoners sometimes tap on walls to communicate in code. Even in silence, messages can be shared!

Some flowers were used as secret messages in the Victorian era. A red rose for love, a yellow one for friendship!

Braille is a tactile writing system that lets blind people read with their fingers. A brilliant way to unlock books!

QR codes store information in black-and-white patterns. A modern-day puzzle that leads to instant info!

The Caesar cipher shifts letters in the alphabet to create a secret message. Julius Caesar was a code-making trendsetter!

Espionage agents use coded radio messages called "numbers stations." No one knows exactly who sends them!

Military submarines use flashing lights to communicate underwater. It's like Morse code but with glow-up energy!

Some banknotes have hidden security features only visible under UV light. Even money has its secrets!

Mind-Bending Math

Zero wasn't used as a number until about 500 CE. What did they do before zero existed?

Pi (π) goes on forever without repeating. It's the tastiest infinite number!

A "magic square" has rows, columns, and diagonals that add up to the same number. It's math that's... magical!

The Fibonacci sequence appears in nature, like sunflower seeds and pinecones. Math is nature's secret pattern!

A googol is a 1 followed by 100 zeros. It's a number so big, it's mind-boggling!

The word "hundred" comes from an Old Norse word meaning "120." Math and history collide!

You can turn any number into the number 4 using a math trick called "Kaprekar's routine." Math has its own kind of magic!

If you shuffle a deck of cards, there's a good chance it's in an order that's never existed before. Infinite possibilities in your hands!

The number 7 is considered lucky in many cultures. It's the universal favorite number!

Roman numerals don't have a symbol for zero. Imagine writing numbers without zero—it's tough!

There are infinite prime numbers. They're the building blocks of math!

Pascal's triangle reveals patterns used in math, probability, and nature. It's a triangle full of secrets!

The concept of negative numbers confused mathematicians for centuries. Who likes owing math, anyway?

A mobius strip has only one side. It's a shape that defies logic!

There's a number called Graham's number that's so big, it can't even be written. It's like infinity's cousin!

A palindrome number reads the same forward and backward, like 121. Numbers with symmetry are satisfying!

The number 9 is special in multiplication because its digits always add up to 9. It's math with a cool twist!

Math puzzles like Sudoku have only one solution. It's like solving a number mystery!

Triangles are the strongest shape in geometry. That's why they're used in bridges and buildings!

The number 42 is often joked about as the "answer to life, the universe, and everything." Thank you, Douglas Adams!

Interesting Illusions

The Müller-Lyer illusion makes two equal lines look like different lengths. Your eyes are playing tricks on you!

The Ames room illusion makes people look tiny or gigantic depending on where they stand. It's the ultimate funhouse effect!

In a mirage, hot air bends light to make it look like water is on the road. Nature's way of fooling travelers!

The spinning dancer illusion can appear to spin in two directions. Which way is she going? It depends on your brain!

The "checker shadow illusion" makes two identical colors look different. Shadows are sneaky tricksters!

The Penrose stairs look like an endless loop of steps. It's the staircase that goes nowhere!

The "hollow face illusion" makes a concave face appear convex. Even your brain can't resist flipping it!

An afterimage illusion happens when you stare at something bright and then look away. Your eyes create a ghost image!

The Kanizsa triangle makes your brain see a triangle that isn't really there. Your mind loves to fill in the blanks!

Optical illusions like the "impossible trident" trick your depth perception. Three prongs? Four? Your brain can't decide!

The "café wall illusion" makes straight lines look wavy. Who knew tiles could be so confusing?

Motion illusions make static images appear to move. It's like the picture is alive!

The Ebbinghaus illusion makes identical circles look different in size based on their surroundings. Context changes everything!

Stereograms hide 3D images in patterns. Can you spot the hidden picture?

The "rotating snakes illusion" makes circles appear to spin. Your brain's love of motion is at it again!

Some illusions rely on perspective, like forced perspective in photos. Objects seem bigger or smaller depending on how you frame them!

The "color-changing dress" broke the internet when people couldn't agree on whether it was blue and black or white and gold. An illusion that divided the world!

The "Troxler effect" makes stationary objects fade from view if you stare at them too long. Your brain loves to ignore what it thinks isn't important!

Anamorphic art looks distorted until you view it from the right angle. It's art that plays hide-and-seek!

Shadow illusions can make simple objects look like something entirely different. Shadows can tell their own creative story!

Quirky Quantum

Particles can exist in two places at once, thanks to quantum superposition. It's science's ultimate magic trick!

Quantum entanglement means particles can instantly affect each other, even when far apart. Einstein called it "spooky action at a distance."

Electrons behave like both particles and waves. It's like they can't decide what they want to be!

The Heisenberg Uncertainty Principle says you can't know a particle's exact position and speed at the same time. It's science with built-in mystery!

Quantum tunneling lets particles pass through barriers they shouldn't be able to. It's like ghosts walking through walls!

Schrödinger's cat is a thought experiment about a cat that's both alive and dead. Don't worry—it's just an idea!

Atoms are mostly empty space. If you squished all the atoms in a human, we'd fit in a sugar cube!

Light particles, called photons, have no mass but can travel at the speed of light. They're the fastest travelers in the universe!

Quantum computers use qubits that can be 0, 1, or both at the same time. It's computing on a whole new level!

Particles pop in and out of existence in a vacuum. It's like the universe is constantly creating and deleting things!

Quantum teleportation transfers information, not objects, across distances. It's not Star Trek yet, but it's still cool!

The observer effect means particles behave differently when you're watching. Even particles get shy!

Quantum foam describes the universe on the tiniest scales, constantly bubbling with energy. It's like the fizz in a soda, but for space!

Virtual particles appear and disappear in pairs. They're like the universe's pop-up guests!

At the quantum level, time can seem to flow backward. It's time travel—but only for particles!

Some particles can pass through the entire Earth without interacting. They're like invisible commuters!

The double-slit experiment showed that observing particles changes how they behave. It's the science world's greatest mystery!

Quantum fields exist everywhere and are responsible for all particles. It's like the invisible stage for the universe!

Quantum physics helps explain how the sun creates energy. Even the stars are powered by tiny particles!

Quantum mechanics is the basis for many modern technologies, like lasers and smartphones. You're holding a quantum-powered gadget right now!

Bonus Fact (#1001!)

The unique black and white stripes of a zebra are actually determined at the molecular level by specialized cells called melanocytes. These cells contain molecules of melanin (made up of specific arrangements of atoms) that either produce the black pigment or don't produce any pigment (creating white stripes).

Even more fascinating - scientists discovered that the microscopic spacing and atomic structure of a zebra's black stripes create tiny air currents that help cool the zebra down. The black stripes heat up faster than white ones, creating small-scale air eddies that act like natural air conditioning. So the atomic properties of those melanin molecules don't just create a cool pattern - they literally help keep zebras cool!

More Fun

Want to be featured in our next book?

JOIN THE FREE FACT FACTORY EMAIL LIST!

HOW TO JOIN:

Use the QR code above or visit

sweetpeapress.net/giveawaysnewsletter

Have a parent/guardian fill out the permission form. Start getting amazing facts every month!

SPECIAL NOTE: Have an amazing fact to share? Use the Comments Line on the form to tell us about it. All facts submitted may be considered for future Fact Factory books! You could even be mentioned in our next book!

WHAT YOU'LL GET:

- The Fact Factory Files Newsletter (monthly)
- Amazing giveaways of autographed items, new books, etc.
- First news about upcoming Fact Factory books

SAFETY FIRST!

Remember:

- Always get a parent or guardian's permission
- Never share personal information online

- Let adults help you with email activities

Want even more facts? Follow Sweet Pea Press on social media with your parent's permission:

Facebook: https://www.facebook.com/officialsweetpeapress/

Instagram: @sweet_pea_press

Made in the USA
Monee, IL
04 November 2026

67380492R00115